Something to Think About

Save Mother Nature

M J Charles

Something to Think About

In this book:

This book has only my personal poems, opinions, statements, sayings and quotes.

Author:

M J Charles

ISBN 978-0-473-45079-3

Contents

Why

I have written this book with my poems, opinions, statements, sayings and quotes to try and get many people to think logically because we can't keep on living the way we do with our powerful rich leaders structuring our lives for their benefit, while allowing the human overpopulation to destroy Mother Nature.

Let's face it, we all know that politics and big business are corrupt, so why do we allow it. "Just because politics and big business has been around for two thousand years doesn't mean it has been done the wright way for all of those years". Many people don't know of any other way that can be better because they have been brainwashed to think that life is all about money. This is what I call "The Evil Root of Money Virus" and if we don't do anything about it, there won't be much of a future for the next generation.

If you get rid of money, then you get rid of the powerful rich leaders, politics and big business, which will get rid of most crime. Then you can start to actually solve human problems that are destroying Mother Nature such as, human overpopulation, pollution, ozone layer and health.

When this is done Mother Nature can go back to its design of having all its elements of life working together as one, by caring, sharing and being equal as each other, to make sure that all life is bright and bold, to stand tall in the paradise of Mother Nature's life.

Nature

Save Mother Nature

Mother Nature
has been designed
for the control of all life
Humans should not dominate
with a greed of Mother Nature
they should have a need
of Mother Nature

Natures Voice

We thought that humans
were here, to care for our needs
Turns out, their needs are for greed
to just, look after themselves

I don't think they know
that they carry the evil virus
It directs them forward to money
dragging us all down the dunny

Cats Are So Proud

Proud to be alive, with a beautiful stride
it's a mate she has to sway, before it's too late

After something to strive, it's now time to reside
with not wanting affray, she says bye to her mate

Her cubs born in a hive, and still having to hide
it must me mums' way, that decides their fate

Getting taught to survive, from a mum who has pride
they'll need lots of play, to turn out so great

Wanting them to thrive, she has to be their guide
so they can have their say, for what's on their plate

With young ones to feed, she goes hunting for food
with a stealth of a creep, she gets close to some feet

It is what they need, as their meal comes her way
she has a big leap, and clamps down on her meat

In they come to reap, on the catch of the day
they now start to feast, after finding the best seat

Forests Voice

We were so plentiful
enjoying what came our way
We were a fantastic team
that had lots of spirit

We started to lose our merit
as humans were taking our credit
But we'll keep on fighting
because they will not break our spirit

We are now doing, a lot of grinding
but all our efforts, are not binding
Unfortunately, bad news is coming
we're losing, to the human ferret

Change

It has been done for thousands of years
so everyone goes on with the flow

Who says it's been done wright, for all off those years
so there must be a big change, for Mother Nature to glow

Our House

They roam their territory, out in their jungle
and with plenty too fell, it gives them a smile

It is their mobility, that makes them agile
With a sense of smell, here comes a new arrival

Not being revolutionary, because their home is not vial
they don't have to yell, because that's not their style

Well known in history, to which one is so vital
they should ring their own bell, to make sure they stay primal

Lands Voice

The production of man
has not been good for our lands
We are losing our girth
and that's not good for Planet Earth

The destruction of our lands
which was making us grand
does not help Mother Nature
as our future is turning to sand

Our lands use to be sound
with all elements doing their part
but with the dictatorship of man
they just roll down our heart

A Clear Message

We have beautiful cats, of colour and pride
to us they stand out, but not those they prey

We have amazing bears, of size and fright
to us they stand out, when their meals come their way

We have cuddly koalas, with a teddy bear look
to us they stand out, but not on the highway

We have large gorillas, with a muscular outfit
to us they stand out, when we need their clay

We have giant whales, with a majestic look
to us they stand out, when their meat is pay

We have greedy humans, with money and wealth
to us they stand out, because it's always their way

All life should be equal for Mother Nature's Life

A Bees Life

Bees are drawn to honey
humans are drawn to money
For bees, it's the life they need
for humans, it's the life they greed

Bees might have a problem
to find all their honey
Humans are destroying their land
that makes them more money

Humans greed the land
to plant the root of evil
so, they can have their taste
of honey, that makes them money

Tigers Love to Roam

Out in the jungle, there is a great predator
which is well known, by all of its prey
It sometimes bungles, a hunt of a creature
not one to moan, it goes on with the day

A tigers' life, is all about survival
and they can smell, that of a new arrival
They are fighters, who like to travel
and grow, so they can prevail

Just like humans, they love a pool
because they want, all that's cool
Just like humans, they have to kill
because it's their stomachs, they have to fill

This cat is strong in life, with its size of freight
it has no say in its future, but wants' what's wright
It doesn't want to fall in strive, just to give in to plight
it wants to stay in nature, so it can show its might

Cats R Free Dogs R Not

To do their thing, cats are free
to want to in sight, dogs just glee

In their territory, cats can roam
in their domain, dogs just moan

In the night, cats pursue prey
in the light, dogs just way

Just like wrestling, cats love to rumble
just like pie, dogs are made humble

As they roam, cats are wild
in their yard, dogs are mild

In freedom, cats will wander
in despair, dogs just ponder

To serve our god, cats have slaves
to no whose boss, dogs have masters

The Back Door

What's really going on, in our back yard
we could be gone, if we don't put up a guard

We are not focused, in seeing our future
we're confused, about living with nature

We're like a disease, growing like a big thunder
it's not a breeze, to stop us from going down under

Can we be fare
so that we can care
for all life
to become equal

So Many Blind

I can see, what is happening
and I know, what's going to happen
so why, are so many blind
about trying, to save earths mind

Who Should Be Culled

They tell us, there's too many
of those damaging mammals
They are destroying the land
for us humans to be grand

We tell them, they're not too many
of those amazing animals
It is humans, that are too many
when they take over, all the land

So Proud

Look at their face, they are so proud
and it's their grace, that makes them so loud
It is their place, for us not to crowd
so let us fight, for their land not to be ploughed

There is something going on, in their habitat
it might not be long, before they can't climb a tree
They need their territory, to not live like a rat
this is their jungle, that sets them free

They are independent, and one hell of a cat
so let us be one, so we all can agree
that this is not on, for us to of sat
let's set them free, so they are able to glee

Just like a god, they are king of the jungle
they like to be wild, and not made humble
With their freedom, they love to roam
so let them be free, in their great home

Seeing Is Believing

Why are we blind to what we see
it's not hard to open our eyes
it's in our yard for all to see

Why don't we look with all our eyes
to make sure Mother Nature survives

Beautiful Cats

They are so grateful, for the land they share
with boundaries to prospect, is for them to desire

They are so graceful, with a spring of a hare
and with their power aspect, is for everyone, to admire

They are so beautiful, in what they wear
and with a mate to select, their breed, will not expire

They are so playful, and something to fear
With cubs to protect, their strength, will inspire

When Opposites Attract

The earth is steady, with a north and south pole
with positives and negatives, for life on earth

This is a cycle, that works very well
interfere with its structure, and it will turn to disaster

The creatures in nature, have a vital role
with predators and prey, it's time to give birth

This is a routine, that lets them excel
interfere with their pasture, and we all go down faster

**Perhaps we all need to go to Specsavers
and get a Mother Nature reality check**

Whose Caretaking

Humans are supposed to be
the most intelligent animal on earth
so why do they have
the inability, to look after planet earth

Attitude of Need

All humans have a need, to feel good and be happy
but humans need of greed, is destroying earths granite

Mother Nature has a need, to feel good and be happy
but its turning to weed, and feels very unhappy

Its death is short, with a dying nature
no one seems to give a dam, about destroying our planet

Where is the caring attitude, to get rid of this evil virus
We need to find an anti-dote, to cure Mother Nature's life

**We cannot think
what is best for myself
my city and my country
We have to think
what is best for
Planet Earth
so we can keep on living
in the paradise of
Mother Nature's life**

Is It for Us

It is not for us, to dictate nature
it is nature, that should dictate us

It is not for us, to make people greedy
it is the greedy, that destroy our future

It is not for us, to make all that money
it is money, that makes as so needy

It Is Time

Time to act—act for life
Life needs help—help us fight
Fight the evil—evil that breeds
Breads the weak—weak we are
Are we used—used to far
Far we go—go with speed
Speed to save—save us now
Now will bring—bring us hope
Hope is caring—caring for all
All are equal—equal as one
One is free—free to be
Be your self—self you believe
Believe in all—all have respect
Respect each other—other are equal
Equal as Nature—Nature is life

Human Overpopulation

Save Mother Nature

The human population has got out of hand
and our leaders don't give a dam
Humans are supposed to be
the care takers of Mother Nature
not the undertakers of Planet Earth

Population

More jobs are created
for us to be employed
so, our heads don't go down

Unemployment goes up
when jobs are scarce
it's, our sanity that goes down

When more buildings go up
to send us to work
it's, our landscape that goes down

The diseases go up
when we are spreading
it's, our immunity that goes down

When crime goes up
because of our density
it's, our life that goes down

When the population goes up
it's the growth they want
for their foot to go down

The accidents go up
because we are crowed
it's, our bodies that go down

When more houses go up
it's the use of more land
for nature to go down

The pollution goes up
when more things are made
it's, our health that goes down

My Thoughts

I say to myself, why do you look to the sky
for a place to supply, us the means to survive

I think to myself, this planet is crowded
so we have to look high, for us to stay alive

I ask myself, why have we become to many
it seems life has been designed, for those who are greedy

I tell myself, we have to do something
for us not to be assigned, a life that is needy

Predicting Growth

Statement

Country and city leaders around the planet make predictions of human growth, so they can plan for the future. The future is not planned for the use of Plant Earths materials, from the human overpopulation disease. Our planet is already running out of material resources and is struggling to grow food for all life, so why are our leaders wanting to breed more humans like a petri-dish. Evan going ECO friendly they are finding it hard to find material resources.

“If our powerful rich leaders choose to ignore the problem of human overpopulation that sits wright in their face Mother Nature’s future is going to turn into a sea of bacteria.”

To Many

Statement

Trump says he's pro-life, but that’s fake news because he only means American human life and he doesn't care about the rest of Mother Nature’s life. We have over a billion of unhealthy humans that we can't look after properly, and Trump wonts more humans we can't look after. I think Trump has realistic interest in Star Wars, as he wonts to produce clone solders, so he can become a Sith Lord. We have to get rid of these evil people.

Logan's Run

Who wants to be culled
when you have turned thirty
It is the fault of our leaders
that we are overpopulated

They have made earth weedy
by creating the greedy
so, we need a solution
to control the population

Human Destruction

Statement

All human problems are a major contribution to the destruction of Mother Nature but the number one problem is human overpopulation. This is very concerning because if we don't take the human overpopulation seriously in the immediate future, life on Planet Earth is doomed. Most of these problems have one thing in common, the root of evil MONEY. Everyone has to change their thinking process and open their eyes to view what is in the media every day and they will find that most of the data is all about MONEY. We have been brainwashed to go forward with the technology of MONEY and are being controlled by our powerful rich leaders. We have to go back and live with the design of Mother Nature's life. "A cleaner's job is just as import as a president's job". Until we get this attitude all life will not be equal.

Urgent Priority

Statement

Why does no one talk about the overpopulation of humans that are destroying Mother Nature. This is a major urgent priority, yet world leaders don't want to talk about it. Everyone talks about a housing crisis but the cause of that is a human overpopulation crisis which is destroying our planet. World leaders fight (compete) with other countries so that they can have the best economy, but they don't care about the human overpopulation disease.

Outside the Square

To those who rule, we all must ask
please think outside the square, for us to stay here

It's humans, that crowd the earth
and it won't be long, before we go up like a bong

They won't more growth, to create more jobs
this spreads more money, for what they call honey

The creation of money, has been bad for our granite
by abusing nature's elements, that destroys our planet

Control Human Overpopulation

We must take the human overpopulation very seriously for the quality of Mother Natures' life. We must have what I call a "tough love policy" so we don't over populate the planet and use up all the material resources. There are so many humans who have many babies with multiple partners, there are parents who struggle to cope with many children, there are too many young parents who aren't mature enough to bring up children, there are parents who discriminate and there are babies born addicted to drugs and alcohol.

--- Everyone should be married if they are to have children
--- A birth limit of two as long as the first birth is one baby
--- You can't have another child with a different partner except for certain circumstance's
--- You have to be 21 to marry
--- Abortion for couples who are not married and for drug and alcohol addicted pregnancies
--- Abortion for unplanned pregnancies

Money
&
Power

Save Mother Nature

We all know
that money
is the root of evil
so why do we have it

Get Rid of the Evil

Money, has created the wealthy
and many, who are unhealthy

Money, has given us disease
with a bacterium, of too many humans

Money, has created the wars
which has made, many sores

Money, has made many do crime
to relieve stress, by having a dime

Money, is not there for many
to have, a good quality life

Money, has been abusing our granite
which can only, destroy our planet

Money, has taken away equality
with no chance, of becoming equal

Money, has taken away life
but life, could have been saved

Money, has taken away respect
to care and be fare, for all life

Money, is destroying our creatures
by bulldozing, Mother Nature

Money, has given us leaders
who greed, for their own need

Money, has created those who are evil
who use and abuse, just like the devil

**Money has created those who are evil
who use and abuse just like the devil**

There is an old saying
wake up and smell the coffee,
but it should say
wake up and see the evil

Money for Nothing

For what is hard, to find in the ground
and something so rare, for people to hound

Why is this shard, worth a pound
it just has a glare, for the rich to be sound

Gold and Silver

Gold, silver, diamonds and rubies are found in the earths granite and they are very hard to find. Someone decided that because of their rarity and glow, they will be used as money. Leaders use them so they can stand out and females use them as jewellery, so they can look beautiful. These leaders have created an obsession of something that is an element of Mother Nature, so they keep everyone interested in looking forward to digging up MONEY.

Gold, silver, diamonds and rubies etc are used for money but no one seems to understand how much damage they have done to Mother Nature. Money seems to be more important than life.

We have a large amount of people who lose money, for a small amount of people to make a lot of money. How can this be fair? We have gambling, share markets, building markets and many other investments that is so prejudice to many people who haven't got much money and they have to struggle with retirement life on an inadequate pension. So, what does this tell you about the Evil Root of MONEY.

Money Gives You Power

There are people, with lots of money
who want more and more
and will deceive and lie, to have their way

There are people, with not much money
who want more and more
and will burgle and rob, to have their way

Don't need the greed of the root evil, MONEY

The Cause

Statement

MONEY is the worst invention because it has turned good people into evil people, just look at the state of Plant Earth today, it's all because of MONEY. All the wars that take over land are because of MONEY. We have so many dictators, corrupt politicians, business and religious people who use a "MONEY go around" so it makes it harder to find out what they are doing. They do this by breeding and manipulating many people to their way of thinking.

The Wealth of Humanity

The cause of the human destruction of Mother Nature is MONEY to which the greedy and powerful have brainwashed everyone to think that life is all about money. So, if you get rid of MONEY, you get rid of politics and business and replace them with community organizations, then you get rid of the powerful greedy. You put in structures like, no pollution, birth limits, making everything recyclable, only make healthy food and drink, electric cars with speed limits and give a lot of land back to nature. These are just a few suggestions. If you get rid of MONEY you get rid of most crime and there will be no more dictators, corruption and war that has been a major part of politics and business since day one.

Politics

Politics is about
trying to solve problems
with the problem

Get rid of the problem
and there are no problems

Our Vote

We select who we want
to make sure we are herd
We give them our say
but who knows, if they vote our way

Politics, Big Business, Religion

Everyone knows that there is a lot of corruption in politics, business and religion, yet no one seems to do anything about it. I believe that because everyone has been brought up with politics, business and religion that they don't know of any other solution for a better governing body.

In education you get taught the ins and outs of politics and business and that it is necessary for a good economy, but you don't get taught how to think about other ways that could be better than politics and business. They don't teach you about what is actually destroying Mother Nature, you get taught their way or the highway, well, were going down the highway and picking up speed to the destruction of Mother Nature because of politics and big business.

Politics, big business and religion has been run by deception, lying and denying. "We have to totally change politics, big business and religion if we are to live long and prosper in the paradise of Mother Nature's life".

Evil Economies

More humans are needed
to have a good economy
So, the powerful have seeded
that of a root, to be greeded

The greed of money
directs everyone forward
to have your taste of honey
so you don't, look backward

Back to Mother Nature
is the only way, for our future
Get rid of this, obnoxious weed
that makes us all, have a need of greed

Wheeling and dealing
can be
hiding and seeking

Collusion

Some businesses collude with each other so they can make more MONEY, which makes other businesses crumble. Wheeling and dealing is done behind closed doors, so they can hide what's really going on. They seek out a lawyer and other partners to make more of a profit and to keep their lies down under.

It's just like gambling where you have a dealer who spins the wheel so you can seek out a winner, not knowing the dealer is hiding a game changing button. A country produces a product and exports the best of that product, which doesn't seem fear for that countries residents to only get second best of their product. MONEY seems to be more important than giving your residents a top-quality product.

The Money Go Around

The rich want, more money and power
so they deceive and lie, to have their way

The poor want, more food and health
so they burgle and rob, to have their way

Big Business

Looking back on history, there are kings of land
who wanted more power, to increase their command

They tell a false story, to take over more sand
to build their empire, for them to be grand

They can't be bothered, to run their affairs
so they bring in the wise, and form a round table

Now they have gathered, to hand out their shares
they make up some lies, to enforce their label

Lying and denying because the truth hurts

Evil Growth

There are many who see that the economy is doing well, so they start believing that the governing body is doing a good job. Economies are about money and the best way to get more money, is to breed more humans. They start deceiving you by saying a wealthy economy is a healthy economy, but that is misleading because nothing can be healthy when the human disease of destruction is spreading so fast.

Stop Making Cents

They can print money, when there's a big dept
For us that's a crime, and we'd end up doing time

It smells of baloney, with this concept
It's for them to shine, so they can have their dime

World Debt

Statement

If all countries are in debt, does that mean they owe money to corporate banks and billionaires. If this is the case then they don't have full control of their countries, so that means their economies are being controlled by the powerful rich.

Are you in control of your life
or is your life been controlled

Political Problem Creations

Statement

Politics allows unhealthy food and drink to be made and sold, so that means they have created unhealthy people. Politics is spending too much money on the health of its people, so it needs to recover costs by putting extra tax on the unhealthy food and drink that they allow to be made and sold that has created unhealthy people. So, this means politics is to blame for creating this problem because this is how they create jobs for the money to go around for a good economy. So complicated but so easy to solve, just don.t allow unhealthy food and drink to be made and don't let them breed more humans for a good economy that is destroying Mother Nature. Politics is about trying to solve problems with the problem. Get rid of the problem and there are no problems, so what's the problem MONEY

Quality of life can only happen when human problems can be solved not just plastered up

Problem Solving

Human problems can be solved, with everyone getting involved
Total change has to be done, or we all, will end up gone

Get rid of this evil money. or it won't end up sunny
It's not much of a life, with us ending up in strive

We need some logical thinkers, and not those who were blinkers
because humans seem to be blind, with all this money on our mind

We have been brain washed to think that life is all about money

Not Solving the Problem

Statement

Yet again the sugary drinks problem comes up and yet again they are only going to try and solve the problem by putting a band-aid on it (Tax}. How about actually solving the problem with a simple solution, don't allow sugary drinks to be made. But the political business collaboration will say, they will lose too much MONEY and MONEY is more important than actually solving problems. So yet again evil has won.

Chain of Problems

Human problems are linked together on a chain of associated links and there are many chains linked together that have been created from the evil root of MONEY. If we don't get rid of this root of evil, Planet Earth will be strangled to death by the human chain of problems.

Powerful People

There are those, who have lots of money
and are happy, to give some away
to those, they want in power

It will make them, a lot more money
and their happy, to have their way
so they can keep on living, in their high tower

Pedal Power

There is a cycle, that we all are riding
it gets bigger and bigger, with all humans spinning
Those who are crafty, will start on the climbing
and let all the others, do all the grinding

Some say that the begging of human life
began with seeing is believing
but no one can see the end of all life

A Dictators World

Statement

Lots of resources will not be available in the future, that is because of MONEY and if MONEY stays around we will run out of all our resources. The powerful rich will get more powerful and richer and the poor will get poorer. The leaders of America, Russia, China, North Korea and the likes, will be dictating the world by playing with their money go around. Politics and businesses has to be totally changed, so we can get rid of these powerful rich dictators. Let's not become the aliens and take over another planet to live on.

Citizens Wrights

There are citizens of countries
that have been dictated to
so, they fight for their wright
to be free, of dictatorship

There are citizens of countries
that have been led by
being swayed and played
to be free, with leadership

The Devils Fuel

The need to have something
turns into the greed of something
and leads to the power of greed
that feeds the root of evil

Trump the Dictator

Statement

There are many leaders (dictators) of countries who don't think about most of their citizens, they just think what is best for themselves. A good example is Donald Trump not allowing children to be with their parents, how inhumane is this. Is MONEY more important than life? "Don't breed humans for an economy, grow humans for a quality of all life".

Dictatorship of Greed

Statement

Planet Earth is facing a dictatorship of greed. Dr Evil wanted to dominate the world and now Donald Trump wants to dominate space (between his ears). Movies are coming true and it might not be long before Donald Trump wants to become a Sith Lord. We need to convert one of his sons to the good side so we can kill of his evil father. May the force be with us.

Playing on Your Mind

We have entertainers who have shows with illusions of trying to make you believe that something can be done. But they have deceived your mind with a misleading trick of science.

Business adds are showing you that they are a caring company, but what they are doing is trying to make as much money as possible by making you believe that they care. Many surveys are asking, do you think a business is caring or not, so by been seen to care you will buy more of their products. Even internal management try to show you they care, so it doesn't come back to bite them when things go wrong.

Excepted a freebie
for something you want
or taken a bribe
to make sure you subscribe

Manipulation

Big business manipulates your thinking by selling a product with a gift of something you would like. It is a form of inducement that they have collaborated with another business so they can both profit from your decision to purchase their product. You have now been sucked in to believing in their brand.

This has become normal practice of the business world these days, so the big companies will push out the smaller ones. These smaller businesses go bust or get bought out, with employment not guaranteed and this is a cycle that goes on for hundreds of years. The only winners from this is the powerful rich.

To Get to the Top

You go to work, to bring in the loot
you put the effort in, to do a good job

The boss is a jerk, in that fancy suit
we feel it's a sin, for him to be a snob

So, booze is your perk, to climb for your fruit
you're not in the bin, when you're the bosses' knob

Don't Suck Up

It is amazing to see so many people suck up to their boss, so they can get what they want. One person told me that I should suck up to get something I want. I've seen people suck up for pay rises, promotions, cheaper schooling for their children and to live on sight for nothing. They suck up by socialising with the boss using alcohol as the biggest influence.

Quality of life can only happen when people don't suck up to get what they want

Gambling Your Life Away

Statement

Gambling is another addiction that prays on the vulnerable because they are desperate to have a better life and for the rich to have a bit of fun. The governing bodies don't get rid of gambling because it creates jobs and makes them more MONEY. They don't really care about the people it affects, as they are too concerned about the greed of a healthy economy. Adds on television for Powerball are made to suck you in to thinking that you could have a luxurious life.

Crime for Jobs

Crime is needed
for a good economy
as it creates many jobs
for everyone, to eat their baloney

When more jobs are created
there is more tax to be collected
So, crime is allowed
because we have a big crowd

Crime helps the economy
to feed, their money go around
for their control, of a powerful command
to make sure their books, are tightly bound

**For someone to make money
many have to lose money**

Crime Pays

Statement

Because the way human society has been structured by the planets governing bodies, there has to be crime. When babies are born, many are not brought up with good morals because their parents, rich to poor, have never been taught about good morals. Depending on who your parents are, were they live and who their peers are, they are breed into crime. Crime is needed for jobs so the governing bodies can have their human structured life of greed for a good economy.

If I was a genie and told the world leaders that I could make a spell that would get rid of crime, would the say yes, that would be great or will they say no, it will cost jobs. My MONEY is on no, this is their thinking process, they need things like crime for their economies. Criminals are who!

Good and Evil

There is no God
there is no Devil
but there is, good and evil
Unfortunately, a good thought
has been used and abused
by those, who are evil

Money has created stress
and stress has created
the need of greed

A Good Thought Gone Wrong

Let's take Facebook as an example, the founder had a thought of making MONEY using a social network platform and when he got it up and running, he couldn't believe how successful it would be, so he kept on developing it, not realising that there could be issues with privacy laws.

So many people are using Facebook and are obsessed with this social media platform, not realising their personal information is been used for big business, crime and bullying. They keep on using it because it's a great way of being kept up to date with friend's movements, but they don't know who or how others are using and abusing your personal information.

There are many big business leaders who want to make more MONEY, so when they can get a lot of data so easily and for free, they will go with speed to gather, analyse and use this data, so they can build their money empire. Criminals see this information and use it to commit burglaries and bullies will use it for harassment.

Seeing is believing
or is it
deceiving by misleading

Look for The Truth

Seeing is believing, but can be deceiving
Thinking is achieving, but can be misleading
Wheeling and dealing, can be hiding and seeking

Hide and Seek

Those who want, to climb the ladder
it's the ride to the top, that makes them sower

They don't want, a broken ladder
so they hide and prey, that their lies are under cover

They do want, a big ladder
for more of that power, to stay in their tower

All we want, is a faulty ladder
to tear down their wallpaper, and watch them cower

Deception

I got conned by someone who I thought was a friend because it seemed he had a good moral character. He got promoted as a headmaster of very prominent high school and part of his job was to make sure the school had a good name in society. He started to employ teachers that he was mates with and some of these teachers had bad reputations. He would give his teaching mates more privileges than others and many teachers were unhappy with this. There were a number of major incidents that made the paper, that he got away with because of the old boys' network could keep some problems out of the paper.

Decision Making

If I told the world leaders
that I could make a spell
that could wipe out crime

Will they say yes that would be great
or will they say no that will cost jobs

Truth Hurts

Are we thinking, about money
or do we think, about honey

Are we thinking, about our health
or do we think, about our wealth

Are we thinking, of the needy
or do we think, for the greedy

Are we thinking, for all life
or do we think, just for your own life

Do you think for yourself, or is your thinking used by others, to get a head of themselves

Total Change

We can't keep living with the human destruction of Mother Nature, just because our so called rich and powerful leaders don't want all life to be equal. Politics, business and religion have been around for some two thousand years and the natural environment has been eroding faster over those years. We must somehow band together and totally change politics and big business and get rid of the rich and powerful leaders who have brain washed everyone into believing that life is all about money. This is why we have to get rid of the root of evil, MONEY.

Do We Worry

For those, who do all the work
do worry, they will tax you

For those, who yearly work
don't worry, you might get some tax back

For those, who live in a house
do worry, it will cost you

For those, who can't work
don't worry, you'll get a hand out

For those, who play their cards wright
do worry, it will cost you

For those, who are in jail
don't worry, they will feed you

$1 World Wide

Statement

Q. Is every country in dept. If so and we don't get rid of MONEY, why not wipe all countries dept and make a dollar one dollar for all countries. If this is not done, then MONEY is more important than life.

Ironic

Statement

There is advertising which says buy NZ made products. ----"Well if you want New Zealanders to buy their own products, then don't import foreign products".

Respect

all Life

Save Mother Nature

Quality of life can only happen
when everyone thinks that no person
is not worthier than anyone else

Except Who They Are

When we are born and starting to grow
we start crawling to find our feet
When we are able and wanting to know
you see our character starting to beet

It' now our innocents you have to protect
you have to teach us wright from wrong
It's our deference you have to project
so, we can learn to be bright and strong

In our youth, you except who we are
as it is us, who are finding our way
We don't want to be someone afar
we want to feel needed, to have our say

When it gets harder, to climb the ladder
that's when good mates, can help you achieve
They help you step up and be louder
as they are there, for you to believe

Discrimination is not very nice
for us, who struggle in life
We find it hard, to roll the dice
and not end up, in strive

It is hard to live life with a mental illness
you feel out of place, for what they call normal
All we want, is respect and kindness
to save our grace, and not feel ad normal

**Everyone should be caring
about the care to be equal**

Life Savers

It seems like no one cares
to what's really going on
Everyone's vision, seems to be impaired
so, we all need to go to Specsavers
or we all, will end up wearing life savers

Life Has Meaning

For life to have meaning
we must work with all elements of life
and be one team on Planet Earth
to care to share to help each other
so that all life is equal and bright and bold
so, we can all live long and prosper
in the paradise of Mother Nature

What's More Important

There is a drug that could save your life, but it costs hundreds of thousands of dollars, which you cannot afford or raise, so you end up dying. How callus, cruel and inhumane is this for these powerful rich leaders to not give you a lifesaving drug because you can't afford it. That is making a statement that money is more important than life. How is Mother Nature going to cope if this is the attitude of our leaders.

Everyone should have a good quality of life not just those who have the power to lead

Power of One

This is our place
that we are destroying
so why are we floundering
because it won't be long
before we are gone

The human race
has to come together
and work with each other
to be the power of one
to cure what we have done

Power Lead

Life on earth, has been dominated
by those, who lead for greed

Life on earth, should be equal
for all life, to not have a greed

**Quality of life
can only happen
when you think of others
rather than yourself**

Good and evil has basically started from day one and most humans accept it as a normal routine of life and that is not fare for all of Mother Nature's life. No one is saying that this structure is so wrong and not logical. If we can't be fare, caring and equal as Mother Nature has been designed for, then how can we accept this routine of human life.

A Just Cause

Just because, a President is important
shouldn't mean
that a cleaner, is not as important

Just because, someone is brainier than others
shouldn’t mean
that they are, more significant than others

Just because, someone has more skills than others
shouldn’t mean
that they are, more important than others

Just because, someone wears a suit
shouldn’t mean
that others, would wear one

Just because, someone has a desk job
shouldn’t mean
that others, would want one

Just because, many struggle in life
shouldn’t mean
that just a few, have a luxurious life

A cleaner’s job
is just as important as a president’s job
Until we get this attitude no one will be equal

A cake is made to celebrate a headmaster’s birthday, but there are no cakes made to celebrate any other staff members birthday. ------ "A headmaster is no more important than a staff member”.
Why is the naming of a prime minister’s baby more important than anyone else’s naming of their babies?

To Succeed in Life

To be successful at what you do
you've got to have the wright attitude
With success there is always failure
so, you can learn from ones' aptitude

When you compete for juicy steaks
there always could be, a slight defect
Let's face it, we all make mistakes
as no one can be, that perfect

Life may grind you down, on a train you want to ride
when it's not going your way, you don't want a big slide
Motivation is on the cards, so you're not going to hide
instead of falling down, you'd rather have lots of pride

To succeed in life, you have to have good mates
they are there when needed, not on skates
To respect those around you, you must find their fate
they help out when faded, not there for bate
To feed their life, you must appreciate
that they are not shaded, by having good mates

Accept and respect that we all are different

Everyone is Different

Respecting those who are different can be difficult because of misinformation and lack of knowledge about a person's background, which could be, social needs, family problems or a mental illness. There are many situations were people who know that someone has a mental illness, will manipulate and use and abuse them for their own benefit. Then there are people who think you look and act like a weirdo, so they try to humiliate you. There are some who don't like you for the color of your hair and for the fact that you don't seem to be intelligent and have a weird appearance.

Their Mentor

At the start, you plant a seed
then you have to prepare
to give them food and water
and a lot of love and care

As they grow, and except who they are
you must show, that you can be fare
as you know, how far to set the bar
you must flow, to find their gear

Give praise, for doing well
for those in the groove
Give advice, when things don't jell
because your there, to help them soothe

Be meaningful, for them to excel
because they do want, to improve
Encourage them to ring, their own bell
and be helpful, for those on the move

For the commitment to manage
you have to believe in their deed
and give them the wright language
so, they feel someone of need

To get respect you must respect others

A Childs Welfare

Is it wright to have a baby you don't want (for certain reasons} or is it wrong to have a baby {or many} that you do want. It is the child's wealth fare we have to protect, as there are so many unwanted babies on this planet and to many bad parents. "Quality of life, not life without quality"

They are Not You

You want to have kids, to fill in your life
you teach them your way, and how to play

Who knows if your wright, in what you show them
it has always been done, so, you have won

They are all different, and not that much like you
treat them for who they are, not like someone in your car

Quality of life
can only happen
when you have parents
that really care

Parental Training

The statistics say that 50% of marriages end up with divorce, so if you put my "tough love policy" in place couples will get to know their partner a lot better for a relationship or not, so it's less likely to have an impact on their children. We have so many bad parents out there, and it doesn't matter whether they are rich, average or poor our lives have been controlled by the power of greed. Most of the blame should go on how society has been controlled by our so-called leaders. You can see what bad parenting has done, it has produced many youth crime, alcohol, illegal drugs, general health, mental health and moral problems. We must make it compulsory for expecting parents to take parenting lessons, as this is a major factor for quality of life. There are babies born and growing up with discrimination by their parents because of their hair colour, their intelligences, their appearance, their sporting ability or not knowing they have a mental illness.

Use and Abuse

There are a few of those, who enter talent shows
because they want to be, one of those pros

Those who do compete, range from young to old
and have the strength, to be bright and bold

Some people start to think, I don't like the look of you
but looks are deceiving, to not bring out a boo

Then out of the blue, you hear a great voice or two
and you think to yourself, phew, am I a bent screw

Being amazed at your talent, they give you a prize
so now it's time, for you to stand up and rise

You need some help, to create a new life
so you listen to the experts, who know how to strive

They say your looks, need a big make up
so we can all feed, our treasure cup

Or what we good do, is just use your voice
with someone else's looks, would be our clients' choice

Because the beautiful sell, for them to feast
they are allowed to use, beauty and the beast

**Those who are evil
will use a good thought
to create their MONEY
to live a life of luxury**

Getting Involved

You go to a night club, to see what you can find
but it's the alcohol, that's on your mind

You look for someone to love, so you bring out the charm
with what makes you merry, they'll be in your palm

You have managed to score, by drinking all that booze
you are now a couple, but who knows if we'll fuse

Evil Bullies

Bullies are not born, they have been created
by those who don't care, about how to be fare

Bullies see something wrong, when they look at you
they think what a freak, I'm going to abuse you

Bullies think their strong, when they look in a mirror
they think I'm not week, I'm much stronger than you

Bullies like to manipulate, by using their power
and they try to punctuate, to make you cower

Bullies like to humiliate, by making you a bower
and they like to dominate, to make you feel sower

Bullies have a problem, of not having good morals
they need to be taught, how to care, share and be fare

**Quality of life can only happen
when people respect you for who you are**

Coping with Life

Life, has been made stressful
by those, who rule the roost
It has become, so complicated
for all life, to understand

There are many, who find a way
to cope with life, as we know it
They find something, to ease the pain
so their life can roll on, day by day

Religion can be, your anti-depressant
or lots of alcohol, will make you forget

Coffee is handy, to help you relax
or you eat candy, to boost your energy

Junk food is on the cards, to comfort you
and illegal drugs are taken, to fly with a high

Appearance can be changed, so you can fit in
or change who you are, so others can respect you

Pets companionship, is a great relationship
to care for each other, in the life of Mother Nature

**Be yourself
self you believe
and don't be anyone else
as it is you, who has to achieve**

Recycling

I thought that when we started to recycle everything, it meant that recycling would be done in the country of origin, but most of it goes to China. China can't handle all these recyclables any more so why can't every country recycle their own recyclables and make only recyclable products, which would considerably reduce land fill rubbish. This does not happen because of MONEY. We should try and make all products recyclable.

The Anti-dote

Just a few suggestions;

---No money/no crime

---I.D. card for items needed i.e. groceries, household items

---Everyone still works 40 hrs a week

---Everything made recyclable

---Recyclable small bins on fence, regular pic-up

---Replace politics with community governing bodies

---Merge similar businesses as organisations

---Only make healthy food and drink

---Control human population

---Renewable energy only

---Limit on alcohol consumption

---Get rid of crumped up living

---No more high towers

---Give back animal habitations

---Control human population

---Get rid of the root of evil MONEY

Fiction or Non-Fiction

I watch a movie, about planets and space
and I think to myself, that could be our place

Movies are coming true, and it won't be long
before we end up in stew, because our planets not strong

Our leaders are planning, to take over a planet
so why are we drowning. on earths granite

Go Back

Just like a book, you always know the ending
as it has, a good and happy feeling
Just like life, there is always a beginning
not knowing, if the devil is dealing

You buy into a book
with a good and happy ending
You won't buy into a book
if it's bad and earth is pending

Going forward, you are given an answer
Going back, you might find the answer

Do you think for yourself
or is your thinking used by others
to get ahead of themselves

Health

Save Mother Nature

**Most humans who are unhealthy
have not been given a choice
on how to be healthy**

Anti-Depressants

Why do people smoke cigarettes
because they want to socialize
or they just want Nicorette

Why do people take drugs
because they want to socialize
or they just want the shrugs

Why do people drink coffee
because they want to socialize
or they just want an apostrophe

Why do people were fashion
because they want to socialize
or they just want compromise

Why do people drink alcohol
because they want to socialize
and make a fool of themselves

Those who worry (stress) about general life feel that they need something to feel good and be happy, unfortunately that thing can be an addiction of greed. This addiction to feeling good and being happy, to relieve the stress of life is what I call an anti-depressant and this could be, change of appearance and attitude, crime, alcohol, illegal drugs, cigarettes, junk food, sugary products, religion and coffee. These products are a must for our leaders as it makes them more money for their money go around to spin faster. This is why they have done or are thinking of doing, is to put a tax on these products instead of having restrictions or stop them from being made.

**Why is life all about MONEY
when our working life gives us stress**

Booze You Lose

You watch the news
and see all those louts
who drink and abuse
by making fools of themselves

To fix their spews
so, they don't go south
let's not sell them booze
to feed their mouth

Drunken Idiots

We have to be thinking, about our alcohol drinking
for those who get drunk, they can't handle their junk

Our leaders let it slide, to save their hide
let's get rid of alcohol, to not become, an asshole

Booze you lose
by making a fool of yourself

To Many Drunks

This is another problem that has got out of hand all around the world. Hospitals and accident and emergency departments are over whelmed with patients in the weekends and holidays. There is a culture mainly taught by parents to their children that its ok to drink a lot of alcohol and make a fool of yourself. There are two reasons for this, 1, parents aren't taught how to be good parents, 2, alcohol is allowed to be made and sold with no limitations by the governing bodies. Everyone who drinks alcohol uses it to relieve stress of a structured life.

Costly Health

In the media, we are told
that lots of us, have too much guts

What we're eating, is not the best
and what we're drinking, makes us a pest

Its healthy food, that's priced to the hilt
chop down the price, so we don't feel the guilt

**Unhealthy food is sold
so why not be bold**

Comfit Food for Stress

Statement

Junk food products can be addictive because they are a good way to relieve stress. For many people junk food is a quick fix because of a hard day of work and general stress of life from a structured society. The smell of junk food itself gives you cravings, but you know it's unhealthy and this is another reason why it is so hard to give up.

Sugar products are very similar, with an instant calmness which triggers a craving of more, straight away. Sugar is a fast form of energy and when you add caffeine it will give you a more focussed long-lasting boost of energy. Many people use energy drinks for working long hours because if they don't stay focussed it could cost life's. There are also other jobs that require extreme concentration to not miss out on making more MONEY.

"Get rid of money and then you will get rid of all these bad products and then you can create a structure that is equal and fare for all life, then Mother Nature will be the anti-dote."

Take You Bake

You eat your cake, and have a tag
of that you take, of an illegal drug

Your body will bake, and not feel snug
you may not wake, so who's the mug

Smoke You Choke

You see them puff, of that they smoke
they think their tough, for a manly bloke

We've all had enough, of that we choke
so let's not bluff, or we all will croak

Speed You Bleed

They make them fast, so why not speed
there not coming last, it's what they need

Nurses fix a cast, to stop them bleed
let's have a blast, at what they greed

Cars are made fast
so why not speed and bleed

Traffic Stress

Lately there has been a lot of media about traffic incidents. People are running late for work, dropping of children, an activity or have a time schedule to meet. Then there are people who think they can drive fast cars, that are allowed to be made and sold to the general public. Then we have road rage because people who cause an incident don't think about others on the road, as their mind is working on a stress problem. The people who receive an incident mainly start stressing because it will cost them time and MONEY.

So Lazy

You pull up and wait, for that car to exit
so you can be near, to enter your store

You park in front, of a chain or gate
just because your rear, has ended up late

We take the lazy option
when given a choice
walk upstairs or ride the escalator
it's our health that says see you later

We take the crazy option
when given a choice
park legally or be a handicap
everyone is thinking, you weedy sap

Attitude Counts

To be good at what you do
you must have the wright attitude
If you put the effort in
you will go to the wright altitude

To have the skills for you to improve
you do the training that helps you move
It's your mates who will take you afar
to show you the gate to make you a star

If you have the wright attitude
and put the effort in
you will enjoy of that what you do

Cover Up Fashion

Why not grow a goatee or a beard
if you don't like the look of your face

Why not were those three-quarter shorts
if you don't like the look of your legs

Why not were no belt with your pants
if you don't like the look of your rear end

Why not put those studs on your body
if you don't like what you were born with

Why not put tattoos on your body
if you don't like the look of your skin

Why don't you like yourself
when you see something wrong in the mirror

Comes and Goes

Those companies come along, when the markets wright
the profit goes away, when books aren't sound

Our jobs come along, when they want height
the work goes away, because of the pound

Our life's come along, by day and night
our freedom goes away, when we are bound

Those problems come along, for us to fight
your health goes away, with all your mite

**How can we be healthy, when unhealthy products
are made and sold, for a good economy**

Stressful Students

You study hard, to get good grads
and with all that stress, you have made the grade

When you have finished, you bring out a sigh
and then you plan, to party with the clan

With all those drinks, you'll get rid of stress
but when you wake, you'll feel so fake

Made in the USA
Monee, IL
04 November 2026

67384591R00036